I0798591

How Animals Evolved

HOW BIRDS EVOLVED

by Elizabeth Andrews

DiscoverRoo
An Imprint of Pop!
popbooksonline.com

WELCOME TO DiscoverRoo!

This book is filled with videos, puzzles, games, and more! Scan the QR codes* while you read, or visit the website below to make this book pop.

popbooksonline.com/ev-birds

abdobooks.com
Published by Pop!, a division of ABDO, PO Box 398166, Minneapolis, Minnesota 55439. Copyright © 2024 by Abdo Consulting Group, Inc. International copyrights reserved in all countries. No part of this book may be reproduced in any form without written permission from the publisher. DiscoverRoo™ is a trademark and logo of Pop!.

Printed in the United States of America, North Mankato, Minnesota.
102023
012024

THIS BOOK CONTAINS RECYCLED MATERIALS

Cover Photo: Shutterstock
Interior Photos: Shutterstock, Getty Images, Science Source, Wikimedia Commons
Editor: Grace Hansen
Series Designer: Laura Graphenteen

Library of Congress Control Number: 2023939058

Publisher's Cataloging-in-Publication Data
Names: Andrews, Elizabeth, author.
Title: How birds evolved / by Elizabeth Andrews
Description: Minneapolis, Minnesota : Pop!, 2024 | Series: How animals evolved | Includes online resources and index
Identifiers: ISBN 9781098245429 (lib. bdg.) | ISBN 9781098245986 (ebook)
Subjects: LCSH: Evolution--Juvenile literature. | Birds--Juvenile literature. | Birds--Evolution--Juvenile literature. | Biology--Juvenile literature.
Classification: DDC 591.38--dc23

*Scanning QR codes requires a web-enabled smart device with a QR code reader app and a camera.

TABLE OF CONTENTS

CHAPTER 1

WHAT IS EVOLUTION?

Evolution is the scientific **theory** that all life on earth has a common ancestor that lived billions of years ago. Living things **developed** from earlier versions of that ancestor. The changes that happened took millions of years.

WATCH A VIDEO HERE!

The scientist Charles Darwin did lots of research on evolution.

DID YOU KNOW?

Almost all scientists believe in evolution because it is testable.

All dinosaurs were born from eggs.

Evolution happens because of natural selection. This process occurs when one individual of a **species** has different **traits** from others. Sometimes the differences make surviving easier for the individual. As they survive, they pass on their traits to their young. Over time, more and more individuals of the

species are born with the new and helpful trait until it is normal.

Evolution is how fish eventually moved onto land. It is how birds began to fly and mammals started giving birth to live young. It is how humans began walking on two feet.

Going from walking on the ground to flying took hundreds of millions of years.

CHAPTER 2

WHAT ARE BIRDS?

Birds are the only animals with feathers and wings. They come in different shapes and sizes. Smaller birds have short wings that help them fly through plants and trees. Larger birds have long wings that help them glide in the air.

LEARN MORE HERE!

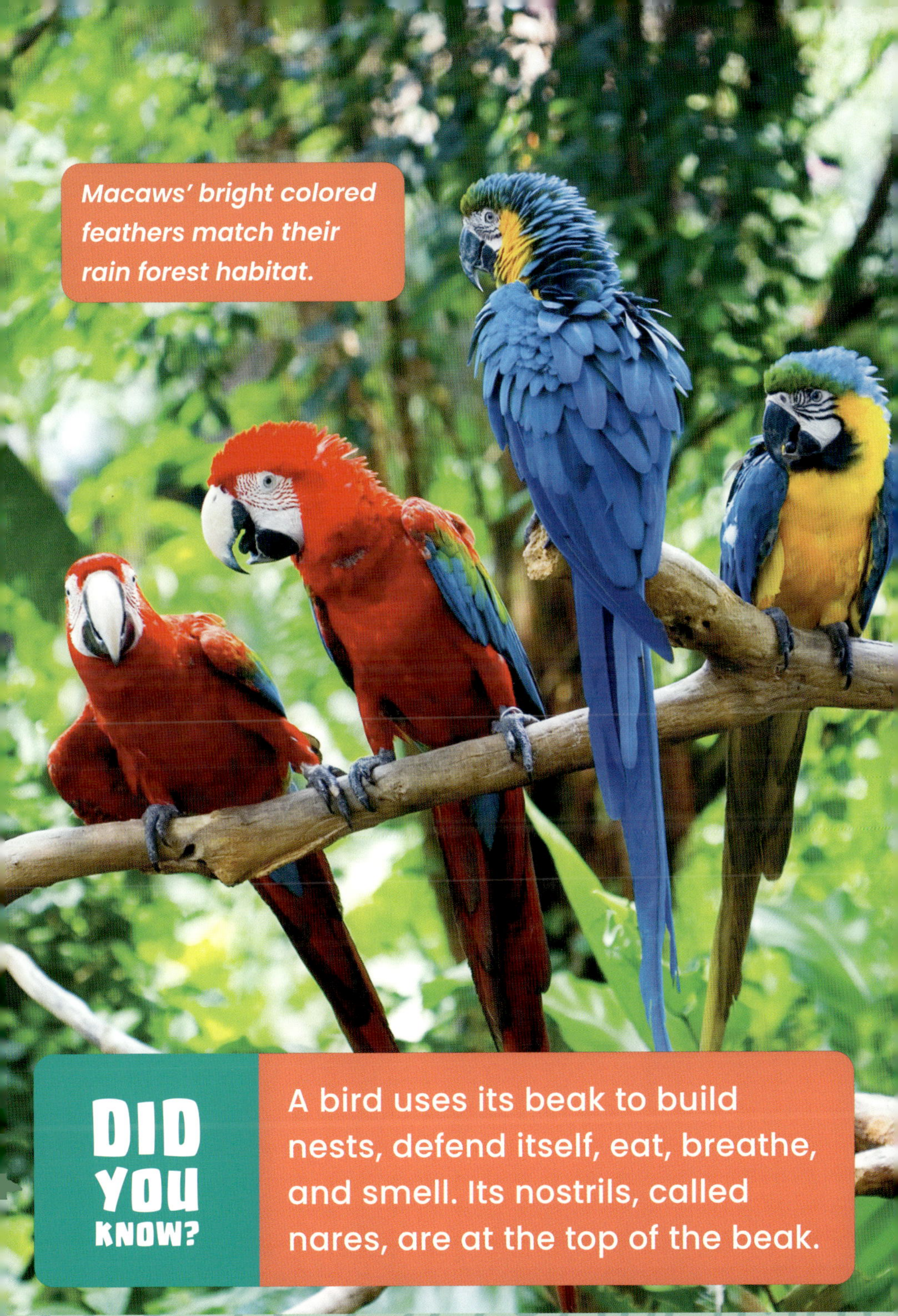

Macaws' bright colored feathers match their rain forest habitat.

DID YOU KNOW?

A bird uses its beak to build nests, defend itself, eat, breathe, and smell. Its nostrils, called nares, are at the top of the beak.

Penguins keep eggs on their feet and cover the eggs with a warm fold of skin.

Birds are warm blooded. Their body temperature stays steady even in cold weather. Birds do not give birth to live young. Instead, they lay hard-shelled eggs in all types of nests.

The number of eggs a bird lays depends on the species. The number of eggs laid by a bird is called a clutch.

There are more than 10,000 **species** of birds. All these species are organized into 40 orders. Orders are categories, such as raptors, waterfowl, perching birds, and flightless birds.

Male birds are usually more colorful than female birds.

A peacock-sized dinosaur called Caudipteryx lived 125 million years ago (MYA). It had tail feathers that look somewhat like a peacock tail.

CHAPTER 3

EARLY BIRDS

There are two ways birds may have **developed** flight. The first began with a feathered ancestor that lived in a tree. It would glide to the ground. This animal learned that flapping its limbs helped it stay in the air longer. The second was a

EXPLORE LINKS HERE!

small animal that lived on the ground. It learned to flap its feathered limbs to catch **prey** in the air.

Deinonychus could grow to 11 feet (3.4m) tall and weigh 160 pounds (73kg).

DID YOU KNOW?

The group of dinosaurs called theropods are ancestors of birds.

EVOLUTION OF FEATHERS

Bird feathers changed in specific ways over hundreds of millions of years. Each stage had a better designed feather than the last.

STAGE 1

single filament

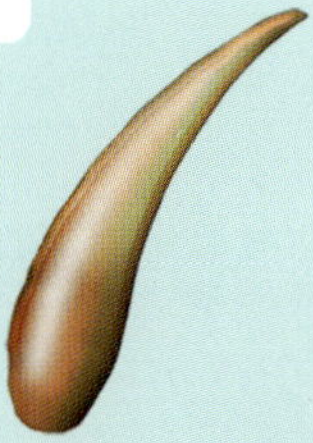

STAGE 2

tuft

STAGE 3

central shaft with barbs

STAGE 4

central shaft with barbs hooked together

STAGE 5

off-center shaft with barbs hooked together

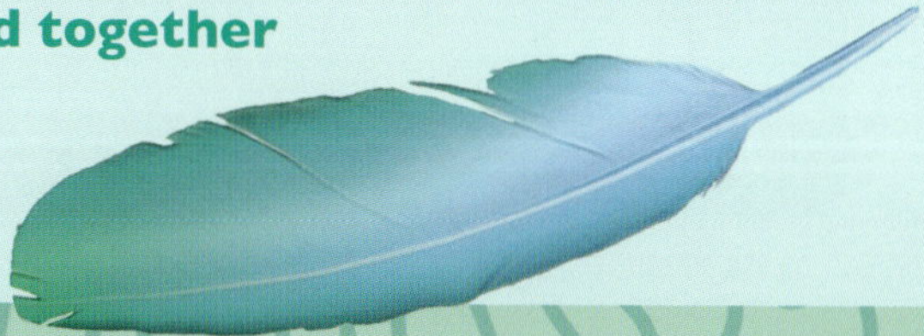

Early in the evolution of birds, specific and important changes happened. First, theropod dinosaurs got smaller. Their bones became **hollow**. One digit on their clawed feet stopped supporting the dinosaurs' weight. Their wrists became more movable to help with holding onto branches.

LET'S TALK ABOUT FEATHERS

The original feather-like structures that appeared in dinosaurs were not like the feathers on birds today. Modern feathers have **vanes** of barbs that zip together and attach to a central shaft. The vanes of dinosaur feathers were made of short flexible barbs that didn't stick together. Instead of a sleek feathered look, dinosaurs looked scruffy and fluffy.

Shoulder bones got bigger to support more muscles and feathers. Vaned feathers started growing in layers. As the body got lighter, tail feathers got longer. Theropods could balance better as they jumped and glided. Shoulder bones developed which would lead to true wings. One toe completely reversed to allow theropods to perch.

This fossil has the feathers of a bird but the teeth of a dinosaur.

Bird ancestors had sharp claws that helped them climb trees.

Starting around 150 MYA, hollow bones began to develop into shapes that could support more weight in the chest and shoulders. Feathers started growing in patterns that helped airflow. Tail feathers were arranged like fans. All these changes led to flight.

Archaeopteryx ***lived 150 MYA. They had light bones, somewhat advanced feathers, and feet for perching.***

Scientists are still trying to figure out exactly how beaks evolved.

An **asteroid** hit the earth 66 MYA. It killed 75 percent of all **species**. The only dinosaurs left were the ones that could fly. All birds that exist today are related to the flying dinosaurs that survived.

CHAPTER 4

MODERN BIRDS

Scientists rely on special techniques to understand the evolution of birds. Much of bird evolution has been mapped by looking at the **DNA** of modern birds.

COMPLETE AN ACTIVITY HERE!

Flamingos are part of the order of birds that include herons and storks.

Wood ducks have feathers that repel water. The feathers keep the ducks warm and dry.

Nectar is the sweet liquid that plants make.

Hummingbirds are incredible little birds that belong to their own order. They started **developing** 42 MYA. They have tiny feet that can only be used for perching. They cannot walk. Hummingbirds live in North and South America. Each **species** of hummingbird has a beak that evolved to drink nectar from specific flowers.

DID YOU KNOW?

Hummingbird nests are very small. They are often only one inch (2.5cm) wide.

Penguins have a lot of closely-packed, stiff feathers. The feathers keep penguins warm.

The oldest penguin fossil is from between 62 and 60 MYA. It was called *Waimanu manneringi*. It didn't look like a modern penguin. But it was flightless and used its wings for diving. The basal penguin from 50 MYA was the first common penguin ancestor.

It was a great swimmer. Many species of penguins evolved from the basal penguin.

Birds of **prey** are also called raptors. Eagles, hawks, vultures, and owls evolved differently than falcons and kestrels. Raptors have large eyes and strong legs. Their sharp talons and curved beaks are made to grasp, tear at, and eat their prey.

A large factor in the evolution of bird beaks is diet. Beaks are specifically shaped for the food birds eat.

The theropods that are the main ancestors of birds were originally predators. One terrifying ancestor, the terror bird, had a giant head like a raptor. It was huge and did not fly. The largest fossil found of a terror bird is 10 feet (3m) tall.

There were 17 species of terror bird.

Tawny owls have started to appear more reddish as their natural habitat warms and experiences less snow.

Scientists are still learning about the evolution of birds. Since fossils of their **hollow**-boned ancestors are hard to come by, new research techniques are leading the way to new discoveries. The thousands of different bird species today will continue to evolve!

MAKING CONNECTIONS

TEXT-TO-SELF

What part of the evolution of birds surprised you the most? Please explain your answer.

TEXT-TO-TEXT

Have you read any other books about animal evolution? What did the evolution of those animals have in common with bird evolution?

TEXT-TO-WORLD

There are thousands of species of birds in the world. With the help of an adult, look up information about your favorite one. Write a short paragraph about your favorite bird.

GLOSSARY

asteroid — a space rock.

develop — to grow over time.

DNA — short for deoxyribonucleic acid, a material in the body that helps determine what features a living thing will inherit.

hollow — having unfilled space within.

prey — animals hunted for food.

species — a group of living things that are very much alike.

theory — an explanation for how things work and why things happen.

trait — a quality or feature of something.

vane — the web or flat expanded part of a feather.

INDEX

DiscoverRoo!
ONLINE RESOURCES

This book is filled with videos, puzzles, games, and more! Scan the QR codes* while you read, or visit the website below to make this book pop.

popbooksonline.com/ev-birds

*Scanning QR codes requires a web-enabled smart device with a QR code reader app and a camera.